a Calya Journey-Wise book

Pendulorum
the practicum on pendulums

by Catherine L. Avizinis

authorHOUSE™

1663 LIBERTY DRIVE, SUITE 200
BLOOMINGTON, INDIANA 47403
(800) 839-8640
WWW.AUTHORHOUSE.COM

First published by AuthorHouse 10/27/05

ISBN: 1-4208-7669-4 (sc)

Library of Congress Control Number: 2005907111

Printed in the United States of America
Bloomington, Indiana

This book is printed on acid-free paper.

Ago and Away. Ago and Away;
And once upon a time;

We agreed to meet like this, and
let our paths entwine.

That Sun has now arrived at
last, and as you read this page;

With the Mystery of the
Pendulum, I wish you Thyme and
Sage.

                    Calya

Welcome, Dear Seeker,
to the Practicum on Pendulums
or PENDULORUM; a book I hope
you will treasure for a lifetime
and remember to purchase again
in your next lifetime, as well! I
ask that you use your colored
pencils to give this book your
personal touch of color in the
appropriate places which will
be self-evident as we progress.
Thank you.

Look for other Calya
Journey-Wise books to help
you progress towards self-
determination and Sovereignty.
The Daily Workbook teaches
the basics of the Journey-Wise
system, RUNES for SEEKERS,
which will be available next
year, and of course Calya's
Chronicles, the novel that will
wisk you away to the Magickal

Land of Hammeril, where Calya and friends learn of their life's calling, embark on a treacherous journey and become honorable seekers in answering that call.

The information in the Calya Journey-Wise books builds a wonderful, integrated system of self-direction. This helps us, not as some might think, to escape into a fantasy world, but to learn of this present reality and it's very Nature; how all things effect all things and how to particpate in our own life's purpose.

But now to class. A Pendulum, known and used by Ancients and Moderns alike, is a marvelous, highly advanced implement that sends and receives vibrational frequency information. It detects and effects and therefore must be wielded with great honor. Know this well. On your Journey, your Pendulum will become an invaluable tool and a trusted friend and ally in every facet of the meaning of those words. Truth, honor and fairness will be required from you.

Many may have you believe that a pendulum can only answer in "yes" or "no" patterns; "He loves me; he loves me not!" or tell the gender of an unborn child. This may be true for the beginner, O, Seeker, but through this book you will now be taken into the higher levels of the **_PENDULORUM_**

Chapter One
Preparing to Begin

You have come to the End of the way things have been; here's a new time of change; for to End is to Begin.

One does not begin a Journey with out some preparation, Dear Seeker, and so we prepare to begin. Here is a Formula that will set Body/Soul/Spirit aglow. Yes, you must learn it, so here we go!

Formula Dulra
The way of things in Nature

"=" read same as or is equal to

"-----" read leads to or develops into

All = One = Collective Consciousness------Universal Principals = Vibrational Frequencies------Velocity = Light------Color = Patterns = Thoughts--------Beliefs-----Archetypes-------Lessons = Extremes------Balance = the Journey of One which effects All.

There are17 points to our formula, 17 vibratonal frequencies , differentiated by colors and Realms, with which we work in the Calya System. Be sure you have your colored pencils at the ready.

Metaphysics is the study and use of Knowledge and Wisdom beyond that which Sciences have proved, meta=over, beyond. Therefore, it , at times, may be deemed folly, weird or even evil to some. It is not.

Thomas Edison said of electricity that he wasn't sure what it was but that he thought we should use it! The only thing it was used for at first was an 18th century parlor game! (PBS) It made one's hair stand on end; quite amusing and magickal and I'm sure to some considered to be Evil. Today, we can't imagine our lives without it.

Through the millenia, Pendulums have been used to locate and detect everything from water, oil, and gold to nuclear submarines; and with great accuracy. Davinci, Einstein, Chemist Robert Boyle and General Patton are all names associated with the use of Pendulums. As is one name in particular, Foucald.

If you have ever been to a Science museum chances are that you have seen a large apparatus suspended from the ceiling, a Foucald Pendulum, tracing delicate patterns in the sand and proving the rotation of the Earth. Foucald didn't invent the rotation, but simply illustrated it's pattern to us with a pendulum. He didn't invent seconds, minutes hours or degrees, but we now use them to explain time. At first with the swing of a pendulum, then with springs and wheels and now with quartz crystals.

Quartz and other elementals and compounds vibrate at certain frequencies and they do this so accurately, so precisely and so consistently that we can power timepieces, radios (crystals), computers (silicon chips), lasers(light focused through a Ruby). Our Ancients began wearing jewelry based on the energetic output they felt from stones and bones. The whole of Pagan and Earth-based tradition is the deep honor of All things. Their dignity and intristic beauty and worth and the unique facet they fullfill in theAll; the effect of the One on the All elevates all to Sacred.

Radios pickup unseen radio waves being broadcast over great distances. X-rays pass through the body without insult or injury to take pictures of the inside of us. Cameras, cell phones, televisions; they all seemed scary and magickal at first.......and who among us can fully grasp the workings and the enormity of the www.com?   But we accept these things now without question as integrated parts of our daily lives. That is unless there is some malfunction. (My OS pdr jpeg what? Can you Ctscan it and FAX me the info?) Even those who have no computers will still be affected by them, like it or not. Much the same as those who do not understand or just want no part of Divinations or matters of Spirit are still effected by them. They exist, therefore, they vibrate!

Your Pendulum has the Natural Property of All things...it has a vibrational frequency output. This is it's character. It has the ability to detect and receive information from other entities and to then effect the energies around it; just like the bird singing in the tree, the rock washed along the beach, the flower in your garden, my white cat sitting in the window and YOU.

You are a Mystery! Who can prove what life is? Why are we here on this small Blue Wonder of a planet which is oozing with life at every crack and crevice while around us Nothing? What is Spirit? Where does it go after death? We, All Things, effect all other things by being. Elements, Chemicals, Consciousness, Thoughts, Colors, Patterns, Extremes, and Balances. If we begin to think we have Life figured out, categorized and neatly slotted ; we will loose Mystery, Magick, Wonder, Thrill and even Hope for a better World in which there builds Tolerance, Compromise, Balance of Extremes, Comfort and Joy for All.

The Calya Journey-Wise System integrates and presents this information helping us to understand and work with these

causes and effects, these Realms and colors.
We, All Things, make up the Oneness which
is the Collective Consciousness, from which
flows Principals of Continuityof Existence,
Survival, yes, but continuing existence in All
Planes. These Universal needs, wants and
desires; true in all places at all times; when
brought down to the bareset essentials, are
Comfort and Joy.

Everyone, everywhere wants
their life to be Comfortable and happy.
Unfortunately, there are those willing to
sacrafice other's Comfort and Joy to increase
their own. We need to seek Balance. Let
us all join the Journey-Quest of our own
Life; participate in this opportunity to be
like Calya and her Friends, learn of our
Life's Calling, embark on the Treacherous
Journey and become Honorable Seekers in
answering that call. While each day brings
new challenges of self-permitting and self-
control, the Realms will give up their
secrets and lessons and the Power within you
will be Acknowledged, Balanced and Used.

Catherine L. Avizinis

Let us try to live again the Formula Dulra
--the Rule of Nature-- and bring Comfort
and Joy to ourselves and others, our One and
Only Planet Earth and the Universe.

Your Practicum
for Chapter One

Be sure you write out your
answers. It will be important to
re-read them later.

1. Why do you want to learn
a higher level of Pendulum work?

2. What is the vibrational
frequency of all things, called
character

3. What is the Formula
Dulra?

Chapter Two
The Choosing

Was it the Twinkle in his eye;
the blush of color across her
cheek?

I held the Pendulum in my
hand and knew; here was my
own Friend.

Deciding on your Pendulum, Dear
Seeker, is not an easy task; so many colors,
materials, shapes and sizes from which
to choose. And , of course, fair warning

should be given; there does exist a condition known as **"PAD"**, Pendulum Acquisition Disorder.…..["But I need this one for Chakra Balancing, and this one for Full Moon…and this one's name is Binky!"] It is only right and good and true to have an assortment of fine tools at one's fingertips; especially ones that twinkle!!!

The first Pendulums of our Ancestors were no doubt bones and wood and stones found along the path, at the scene of some dramatic occurance or in the Sacred Grove. Through the many centuries of Clinical Trials and Observations, all of our theory and traditions of the Powers and use of Crystals and colors, herbs and oils have been established. Magickal Formulas form around the repeated ritual resulting in the desired effect. If a headache was relieved by burning Lavender and using an Amethyst Pendulum in a clockwise pattern once; it should be tried again. If it works three times, but then the fourth and fifth it doesn't, what were the differing conditions; the moon phase, the weather, the cause of the headaches? One can certainly understand how it would take centuries of trials and errors.

Amethyst works on the Realm of the eyes, ears,nose. It is it's Purple light color frequency which effects this body area… Iron, Aluminum and other things help give Amethyst it's Purple frequency. Different chemicals and elementals give the various gemstones, plants, foods your clothing and thoughts their unique characters, their vibrational frequency signature or output. Purple always will have the attributes of Raw emotions attached to it's power. Issues concerning intuitions, feelings and psychic energies; all the "Claires"; are detected and effected by anything which vibrates at the Purple frequency. Lavender, which has a Purple flower, may help a headache or worsen it because it will effect the amount of enrgy/power in that Purple Realm.

Turn to the Charts on the next pages. Carefully begin to familiarize yourself with the 17 Color Realms, Body Areas, and Attributes; and it's Name, it's Purpose, what it helps to balance and what to use if one is Extremely Excessive. These form the basis of the Calya Journey-Wise System; wherein you will find greater teaching information. If you do not have the <u>Daily Workbook</u>, I would encourage you to purchase one on-line or at your local bookstore.

Now, Class of Dear Seekers, please take out your colored pencils and color in the appropriate hue next to the chart information in whatever pattern seems good to you.....Go ahead, we will wait for you. do this now...

****

Color	Body Area	Attributes
White	Above Crown	Stillness, One, Belonging, All
Pink	Crown	Self-direct, own Realm, dignity
Violet	Brow	Gratitude, calm, realignment,
Purple	Eyes, Ears, Nose	Emotions, psychic, justice, feelings
Indigo	Mouth, Jaw, Occipital	Will, choice, voice imagination
Blue	Throat, Neck	Self-worth, -joy, -measure, -truth
Aqua	Shoulders	Receive without guilt, play
Green	Chest, Arms	Giving, healing, love, acceptance,
Lime	Xiphoid	Digest, Striving, Manifesting
Yellow	Solar Plexus, Stomach	New Awareness, changes, Instinct
Peach	Mid-Abdomen	Self-defense, immune, fears
Orange	Lower Abdomen	Intentions, Wants, Needs, Desires
Red	Pelvis	Creative Life Force, Motives
Burgundy	Thighs	Make a Stand, Honor, Battle
Brown	Knees	Stability in Flexibility
Black	Feet	Journey, Ground, Leave your Mark
Gray	All Over	Retreat, Wounds

Color	Realm	Name	Balances	Excess?
White	Above	Moon Day	Longing	Black
Pink	Crown	Calya	Self Rely	Brown
Violet	Brow	Fahaltryn	Change Mind	Lime
Purple	Eyes	Pryll	Feelings	Yellow
Indigo	Mouth	Daemus	Choices	Peach
Blue	Neck	Kriel	Worth	Orange
Aqua	Should-ers	ly	Work, Play	Burgun-dy
Green	Chest	Tesque	Giving	Red
Lime	Gall	Jaedn	Striving	Violet
Yellow	Plexus	Courtheon	Expand	Purple
Peach	Spleen	Delania	Defense	Indigo
orange	Low Ab	Zaphir	Intent	Blue
Red	Pelvis	Oban	Motive	Green
Burgundy	Thighs	Tolus	Warr-ior	Aqua
Brown	Knees	aMaun	Flexible	Pink
Black	Feet	Dalwynara	Excess	White
Gray	All Over	Brymn	Fatigue	Gray

14

Did you see how each color relates to a specific Body Area and specific aspects of issues? These will be helpful when choosing which of your Pendulums to use; choosing one best suited to the Balancing need.

The best results will be achieved by using a tool made of a natural material; gemstone, crystal or metals. These are all very good conductors of Magnetic fields and Receivors/Transmitors of Magickal Energies. Let us now find you a Pendulum.

Plan a shopping trip. [or shop on line at www.pelhamgraysom.com]. Either way clear your head of stress, take a few deep breaths and focus on the Seeking at hand. Speak to the Universe…"**I am in need of a friend, please; a Pedulum to Guide me.**" Repeat this three times, three times a day for three days. In Sacred Geometry the number three sets forth Creation Energies, you are Creating the Pathway for a Pendulum to enter your life. Expect an answer.

Using the Color chart, decide what the main task of this Pendulum will be; what is up-most in your heart and mind at this time?

1) Gray...Soft Comfortor

2) Black...Sacred Journey Guide

3) Brown...Discerning Investigator

4) Burgundy...Honorable Warrior

5) Red...Creative Fire Keeper

6) Orange...Vocation Instigator

7) Peach...Charming Defender

8) Yelow...Enlightened Teacher

9) Lime...Purposeful Manifestor

10) Green...Loving Healer

11) Aqua...Playful Nurturer

12) Blue...Worthy Joymaker

13) Indigo...Mystickal Teller

14) Purple...Magickal WizeOne

15) Violet...Grateful Mind Changer

16) Pink...Sovereign Supporter

17) White...Mighty Uniter

The metals used with them also have significance.

Silver/pewter=gentle leading,

Gold=powerful starter, Copper=quick conductor. As Quartz is a White frequency, Mighty Uniter, and also Magnifys all other energies, it is sometimes a good place to start.

Then there are the Shapes. The classic 6-sided crystal point, spheres, faceted spheres, eliptoides, Egyptian cut [eliptoid with a ball top], cages, spirals, teardrop.….. and all manner of combinations. Any good shop will allow you to hold and speak to the Pendulums they offer (one of the draw backs to on-line shopping!) assuring their pieces are compatibly matched to their new charges. See if one will readily work with you by holding it over your open palm by the very tip of it's chain. Some come with a charm or fob and some do not; but you can always add one of your own device later. Does it capture the pattern easily? If your right side is your strong side then your right palm should run clockwise and your left should run counter- clockwise and just the opposite if you are left-handed.[

left clockwise and right counter]. Ask the pendulum if it is agreeable to go home with you….back and forth "YES" and side to side "NO". Don't forget a pouch or box as a Keep.

Know full well, one will capture your attention and you will not be able to leave it behind.

Once you are home, and this applies to any time it needs cleansing for they do accumulate Magickal information and Energetic charges, wash it carefully w/ a pure and natural soap under clear running water; a flowing stream or a warm gentle faucet is fine. Rinse well and dry with a natural fiber towel or even pure, soft paper towel. Set on a tray of crushed salt in the Sunshine and then leave it overnight in the Moonlight. You will both feel fresh and clear and sparkling.

But I know full well, Dear Seeker, that as soon as you get alone with your new Pendulum, you will dive into the bag, pull out your new ally and the two of you will be chattering, working , giggling and wondering at each other!!! No harm in that at all…What Joy! Play and show and tell; then later cleanse, sit, and as you are drying it, study.

Being polite…learn the attributes of your friend. What attracted you? It's shape? It's color? Are there fractures and facets? …swirls and twinkles? Is it a heavy presence in your hand or a lightness? Know all these things for it is dear to you. Learn all these things for it is learning of you. It chose you as much as you chose it. Does it feel masculine or feminine or both or neither? Who is this new entity you've brought into your Keep? Make your introductions and politely ask it's name. Some will speak quickly; hear it inside. Sometimes an overture name is given, one that is for you to use as the two of you learn of each other. Others take months to share a name. Please be patient.

Some are bold and bossy, some are shy and delicate, but all must be treated with respect. This is not a prisoner or slave and you are not some cruel task master. It is necessary to learn to speak kindly and always ask for your Pendulum's assistance. Please try to remember not to command or demand; a good rule of thumb when dealing with all and any in this plane or the Other.

Some Pendulums are "HAMS" , eager and willing to perform and do tricks for anyone at a moments notice. Others will be sulky or resentful for being put on display and I'm sure I don't need to tell you who will get the last laugh as your usually co-operative Pendulum quivers and shakes or hangs there dead still!!!

For now, Oh, Worthy Seeker, you have sought out and found your Pendulum. My best wishes to you both!

Your Practicum for Chapter Two

Be sure to write out your answers. It will be important to Re-Read them later.

1) Revieve the Charts until you can say the Color, Body Area and Attributes comfortably.

2) Cleanse you Pendulum and Ask if it prefers to sleep in it's Keep or stay out on the salt rest

3) What is the Formula Dulra? Explain it to someone in words.

Chapter Three
Patterns

In Seasons and Cycles; Ebb
and Flow,
   On, does the Clockwork
Universe Go!

I hope you and your new Pendulum
are doing well; learning and accepting each
others fascinating facets and fractured flaws.
…….and why is it that your Pendulum (and
you yourself) have the ability to detect and

effect other energies? Yes, because you both have the Natural Property of all things; vibrational frequency output, character, your *unique pattern.*

In this chapter we will learn of *Patterns.* When asked properly, a Pendulum will gladly communicate with you through a pre-established pattern code. You must give it a few moments to register and learn the code, then practice a few times with "say, yes" and "say, no" to be sure it is clearly understood.

One of the simplest beginners codes is "to and fro"(towards and away from) for "yes", "left and right" for "no", and circle for "maybe".

Calling your Pendulum by the name you are using, (even a pet name is fine, like Dear Heart, Little One or Boppin), ask if it is willing to work with you right now.

Explain that you would very much like to try the beginner's code as stated above. Clear a Field on a space or table in front of you and holding the charm at the very end of the chain, still the Pendulum, ask the question and wait for the answer to develop into a pattern. Do not rush and try not to move your arm or hand, clear your thoughts, calmly and ask a few "known-answer" questions. "Is it raining outside?" or "Is my hair red?" "Is today Sunday?" What responses do you get? If the two of you are working well so far, ask it something you don't know. "Will Julie visit me today?" Note the answer and check for accuracy later.

Work in this fashion until the two of you develop a good working relationship and the accuracy of the Claire-sensients becomes reliable. There is no hurry, take your time. Request, please; do not demand, thank you.

The next step is to introduce a more complex pattern. The following chart is a map of Hammeril, the Magickal Land where Calya, Dithero, Dourstan, Non and Delania travel on their Journey-Quest in the novel Calya's Chronicles. It is also , Dear Seeker, the color map we use in the Calya Journey-wise system. Take a few moments to read it over with your Pendulum. Note, together, the colors, names and locations of the areas or "Realms". Do you see their correlation to the charts you have previously studied in Chapter Two? Please take time to do this all now and with your trusty colored pencils, color in the Map; it will help you to become familiar with the ley of the land. Be absolutely certain of the placement of each of the colors; don't hurry...this is your Journey-Quest Map...finish it for homework if you need extra time. It is essential that you do this...I will be around to check your work later.

This Map will serve you well , my friend. With this code you will be able to find the color vibrational frequency of people, places, issues and things and,

by applying the Calya Journey-wise
information, better understand their state
of existence; the Energy working in and
around them at that very moment you
Read; and therefore, better understand their
struggle to balance, their lessons to learn.
Of course, this applies firstly to yourself; for,
Dear Seeker, if you dare not look at yourself,
dissolve illusions and desire to balance and
change; living honorably in your own Life
and on your own Quest, then dare not try to
tell others how they should travel. Beware
the Guide who walks backwards! Safe
Passage and Journey-Wise!

****

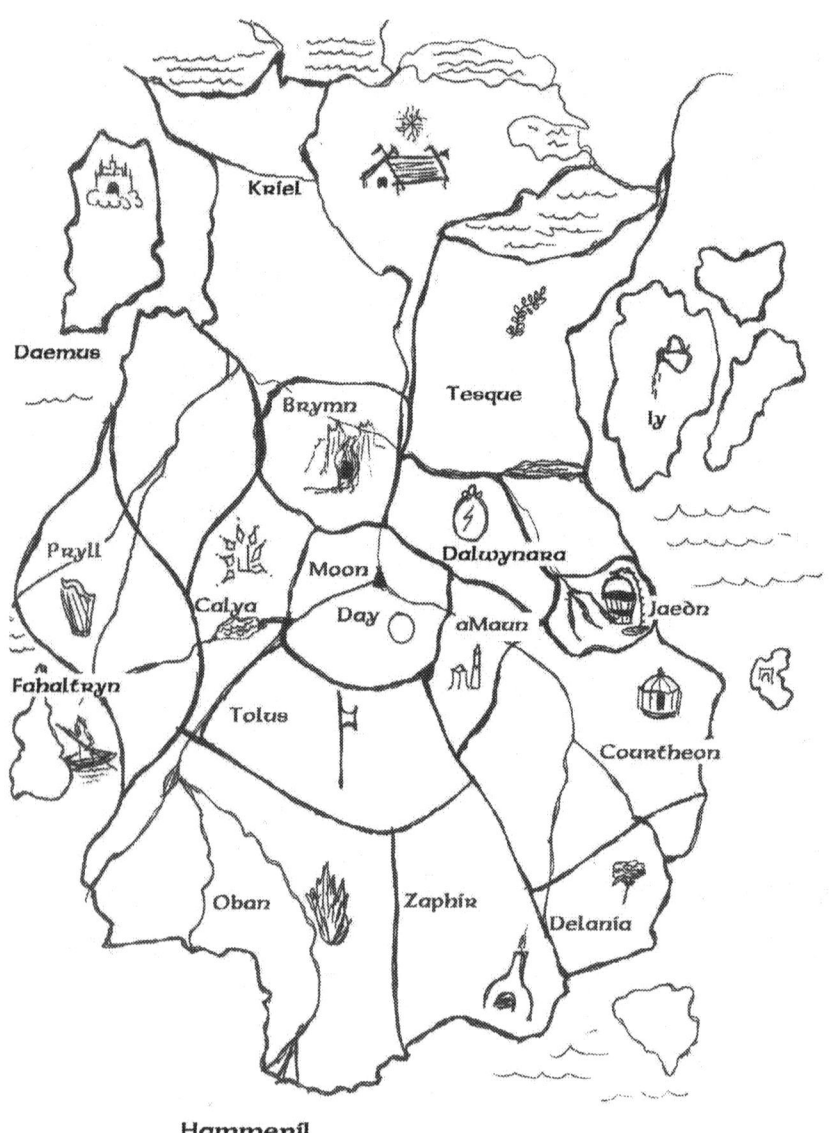

When you are ready and your map is in full bloom of color, place it in front of you on a table or steady surface. This is now your "Field"; the Cosmos into which you will delve Seeking answeres, direction, guidance and hope. Hold the point of your Pendulum over the central Realm of Moon Day, White, and ask,

"Where follows the Path of my Journey-Quest?"

Your Guide will develop into a swing pattern between the Realms you are Seeking to balance this day:

Black/White = your foot path in reality/ your spirit directed path from the Before Time

Brown/Pink=Acknowledging previous agreements and others authority/ your own free will, Sovereign self-directedness

Burgundy/Aqua= Fierce Warrior for just causes/playful-self who needs pampering

Red/Green=Creative fire, ambitions unbounded/healer of all, re-structure

Orange/Blue= True Cosmic authentic self/self here molded by family and society

Peach/Indigo=ability to charm and get own way/ magical connections done honorably

Yellow/Purple=Expansive changes / controlled releases of Power

Lime/Violet=things happening now/ isolation and changing the mind

Gray=retreat, confusion, wounds, rest

Then ask, "Which of these do I **need** for balance?"
The Pendulum will swing to the color that you need to add to your energy field and surroundings that day; through

crystals, essential oils, colors, thoughts in the Chapter work, movements and of course the Releasing Meditation; which give us a physiological response to a mental stimulus causing unwanted or unhealthy patterns to release and wanted, healthy patterns to be reinforced.

Please, seek out the Calya Jouney-Wise Daily Workbook for further information about the Realms and the balancing system. It works and it works every time.

As you become more acquainted with the mysteries of the Color Realms you will no doubt eventually want to ask questions about family, friends, relationships, issues and plans.

"What is the energy surrounding my job promotion?" check the color.

"What is my sister's energy today?" check the color.

"What would be the best action at this time concerning writing a new book?" check the color.

Catherine L. Avizinis

Your life is a series of lessons, extremes of things that seem to be direct opposites. Finding the compromise and balance between these extremes is Love…actions of acceptance, validation, caring compassion for self and all concerned. This Love takes a Heart full of Courage, a mind willing to think and be imaginative, and Hope.

Your Practicum Exercises For Chapter Three

1)Explain to someone the 17 color Realms, using all the Charts at your command.

2) Compose a Rhyming couplet in praise of your Pendulum. (Pendulum sweet, Pendulum fine, I love the way you twinkle and shine!.)

3) Cleanse your Pendulum and give it and yourself repose.

Chapter Four
The Chart

It is now Precious, this Human-kind to me and I shall open wide my heart, so Mote it Be!

You are now ready to move to the next level of Pendulum work. You first entered this class as a Seeker, willing to seek and find for yourself new knowledge and teachings on the Pendulum. It is time now to open and use all that you have learned turning knowledge into wisdom, learning into Practicum and you into a **Keeper of the Realms**; for once you journey past this *marker* and take upon yourself the next steps, you will then carry with you the responsibility to be the Honorable Keeper of one of the secrets of the Realms that very few in this plane know.

For those of you who have completed the suggested tasks in the Practicum, "Hura!" and thank you. You will surely find your progress made easier by walking on the stepping stones set before you. Enlightenment takes discipline and honorable practice. For those of you who have excused yourself from the Practicum Exercises, …tsk, tsk, the road is longest for those who will not walk.

Please bring forth your Pendulum......

"<u>Name</u>" My Pendulum, my dearest Guide............So happy am I to have you at my side..........

Journeying on together, we two,
I will open wide my heart, will you?

It is now time to enter into a deeper trust with your Ally. What is an Ally? One on whom you can count to share information, to back up your position, to "cover your back"; that is, the vulnerable area you cannot easily see. Your Pendulum represents to you a Spirit Guide, a Guardian, an unseen Benefactor. It is one who will help you to understand things you cannot quite grasp otherwise, by means of secret codes between you; one who will confirm or deny your suspicions or feelings keeping you honorable in your dealings with others, especially in areas where you cannot see all of the hidden aspects or information of a situation. Your Ally might fill in the missing pieces keeping you from judging too harshly, acting too rashly or self-negating and not acting at all!

Until now you have communicated with your Ally and Guide by means of a code you have suggested or the Calya Journey- wise map. You will now advance; learning to construct your Pendulums own unique Personal Chart. Now, Seeker, if you have been diligent in your practicum and ask with grateful friendship, your Pendulum will open wide it's heart to you and teach you it's own

Secret Code.

1. Decide with which Pendulum you will work.

2. Gather some art supplies, (white poster board cut to a workable size @ 6"x8", your colored pencils, a ruler, something to trace as a circle or a compass, a No.2 pencil)

3. Address your Pendulum with the "My dearest Guide"… used at the start

of this chapter, then ask if it will please teach you it's Color Code. You may ask it to answer you with a "Yes" or "No" pattern but, believe me, I have never met a Pendulum who was not very eager to share this information about itself. So few are asked!

4. Understand and remember well, this is a **Calya Journey-Wise Realms Keep level** to which you now ascend. Congratulations. We are very proud of you. Once you have the Color Chart you may call yourself truly a **"Keeper of the Realms"**; one who protects and preserves the Ways and Wisdoms of the Realms.

5. With your ruler mark a center point on your board. Using the 17 colors you are going to ask your Pendulum to show you it's own corresponding pattern for each of the colors. Hold the point over the center position you marked and ask "Show me your White Pattern, please."

6. Your Pendulum will begin to move. Wait for it to fully develop the shape and size it wants to show you; straight line, circle, elipse, flower, star, ...then lightly with your No.2 pencil, sketch the pattern it inscribes and mark the flow direction with little arrows.

7. Ask, sketch, color the correct color and shape and size and mark the direction flow with arrows for each of the 17 colors. Always start directly on the center point, take a deep cleansing breath, and relax as you ask politely to be shown the next color pattern!

8. Be accurate, slow, do not hurry this. If a particular pattern is quite small, sketch it in larger scale over to the side, where it will be more easily identified later. Enjoy this wonderful colaboration, the artistry, and remarkable results. No two charts are alike!

9. When you are finished thank and cleanse and Marvel!!!

10. Review your Pendulums Code with the information from the Calya Journey-Wise Charts. **Keep** them together in your Magick Book or special case or Keep. However, You are a Keeper of the Realms now and may wear your Pendulum openly at your waist looped through your belt, around your neck, or pinned to your lapel…always at the ready, Oh Gentle Keeper.

Practicum Exercises
For Chapter Four

Just one this time O,Worthy Keeper…

Define Honor.

Chapter Five
Refinments

Entwined Hearts...My Friend...
My Own;
    Together until we arrived
Home.

    Keeper of the Realms! High
learned are you. Keep yourself
on the Road of Truth, Goodness,
Comfort and Joy for all

concerned. Be not self-deceiving
nor deceptive when dealing with
others. Say what you mean and
mean what you say. Listen, and
hear not what you think the
other is saying but that which
they are truly trying to say.
Be kind always for we are all
tender, wounded, struggling
souls. Seek Comfort and Joy
for yourself; know that you are
tired and hungry and in need
of relief. Know that there are
those who are just too strong
and selfish for you. Yet, also
that there are those whose hand
you need to grasp for their sake
or your own. Your Guide will
show you the Way; your Ally will
guard your back, protecting you
from extremes. Trust your deep
inner self and always be willing
to change for the Harmony of
all, for the Tolerance and the
Peace.

1) Discerning Hand Energy Flow- For someone who is strongly Right handed, the Right hand shows forth a clockwise circular pattern and the Left hand will show forth a counter-clockwise pattern. Test on the up-turned palms.

If there is imbalance, it will show as distorted shapes or even opposite flow direction. This indicates fatigue, unsorted stress, or confusion about issues. Use the Journey-Wise system to balance interiorly then, usung your Pendulum, try to re-instate the proper flow. Hold the point over the center of the palm and swing itin awidecorrect flow direction. Next, grasp the chain at the top near the charm between the thumb and finger and gently slide down towards thePendulum. This will reduce the size of the circle and intensify the flow. Repeat if necessary until the palm picks up the corrected pattern.

2) To help with a cold, Yin condition, congestion, stiffness, constipation, slow

circulation or a depleted Realm pattern; swing the Pendulum in a wide, open, clockwise Yang pattern.

In case of a hot, Yang condition, dryness, inflammation, diarrhea, too open or excessive Realm pattern; swing Pendulum in a small, closed, counter-clockwise Yin pattern.

3) A water pattern is usually a five-ten beat straight line pattern which then alternates with the same number in the opposite direction and then keeps reversing. This is the effect moving water has on the magnetic fields of the rock/earth surrounding it. It also indictes the "Ocean of Emmotions" within in person. Check the Purple Realm balance.

4) A Vortex is a spiral shape or circular Pattern. It acts as a conductor or conduit for shifting, traveling energies. It may indicate a portal or thinness in the Veil between the Planes.

Be cautious and be Honorable. Cast your Protective cirle around you by pointing your power-hand index finger or a wand around the area turning in a clockwise direction, saying three times-

" Only those who wish me well do I invite here nigh.

All others be gone who wish ill or nil, pass by, pass by."

Then open the vortex within the confines of your protective circle with a clockwise pattern; knowing only helpful energies will be able to enter. When you are done conversing say three times,

"Thanks be to you for coming here.

Depart with blessings now, my dear."

As you swing closed the vortex with a counter-clockwise pattern. Release the circle by turning in a counter-clockwise direction.

5) Charting a house or a room- either walk the space or sketch a diagram of it's aspects and dowse over it. Doesaroombeing considered for a study show heavy chaotic energy? Choose another room; preferably

one with strong,straight logical patterns. A business may enjoy a busy traffic pattern at it's door, a room for peaceful meditation or sleep may not.

6) Use the Hammeril Map (your color map) as a power grid chart for creating good energy balance in any area. Mark your Diections, West, North, East, South and decorate according to color and energy and tool. My home, the showroom at Pelham Grayson in Stonington, CT, and a lovely retreat center in Lebanon, CT, are all accomplished in this manner as are the spaces of many students and Journey Guides of the system. It makes for great peace and great vitality as the energies pulse one to the next.

7) Take your Pendulum's readings on different objects and log these in your Book of Shadows and Light......the wooden chair, table salt, your gold ring, a rock from your yard, a piece of Quartz. Keep carefull notes.

8) Experiment with that ring. Test it's pattern over and over. Have someone hide the ring, see if you can find it by dousing, walking with the Pendulum. Try locating it by querie, "Show me Red, please, if it is in the South…" Use your map as a help.

9) Do the same if you or a friend has an ailment. See if you can discern the Yin/Yang of the condition. As you advance in your ability, help to bring the condition to balance with the Journey-wise system. Always and only if the person asks for help, gives their expressed permission and never force your "help" on anyone.

Lastly, Dear Keeper of the Realms

10) Sending Energy into the Astral Plane-When you know your Color for the Day and are doing the Releasing Meditation or Ritual, ask your Guide to help you send out Energies into the Astral Flow. If you are working in Blue send out your Blue into the Rainbow of the Flow to be used where it is needed most; not where you think it is needed. Send it out and offer

it up (see the Ritual pages in the <u>Calya Journey-Wise Daily Workbook</u>) with music, incense, drumming, aromatherapy, song, dance, crystals, colors and the swing of your Pendulum. Freely let it go for the Love of One and Good of All in Comfort and Joy!

May your gentleness stay your hand when your Power you would weild.

May your Power be in your Gentlenss, and your Allies be your Shield.

It has been my Joy to share this Word with you. You have been my Comfort, knowing you are learning the magick of the Pendulum and will share and tell, help and heal; for we must never forget again *The Pendulorum.*

Safe Passage and Journey-Wise,
Calya

About the Author

Catherine L. Avizinis (Calya) lives the magickal, mystical life about which she teaches and writes. The system is the result of her own metaphysical abilities and years of intuitive/psychic counseling.

Journey-guide to many, she is also wife of 29years, mother of three fine sons, artist, musician, Cert. Holistic Aromatherapist, Cert. to teach Legat Russian Ballet, dance and movement, licensed Cosmetologist, author and developer of the Calya system and the Calya Aromatherapy line.